Charles William MacCord

The Teeth of Spur Wheels

Their correct formation in theory and practice

Charles William MacCord

The Teeth of Spur Wheels
Their correct formation in theory and practice

ISBN/EAN: 9783337105679

Printed in Europe, USA, Canada, Australia, Japan

Cover: Foto ©berggeist007 / pixelio.de

More available books at **www.hansebooks.com**

THE

TEETH OF SPUR WHEELS;

THEIR CORRECT FORMATION

IN

THEORY AND PRACTICE.

By PROF. C. W. MacCORD.

HARTFORD, CONN.:
PUBLISHED BY THE PRATT & WHITNEY COMPANY,
Manufacturers of Machinists' Tools, Gun and Sewing Machine Machinery, &c., &c.
1881.

PREFACE.

As a preliminary to the description of the machines for the accurate formation of cutters for spur wheels, given in the second part of this treatise, it seemed appropriate to explain the manner of laying out the teeth. And in the belief that it may be acceptable to many who are interested in the subject, we have endeavored to give, in as simple, clear and brief a manner as possible, the reasons and the proof of every step in the construction. The method of drawing rolled curves by means of tangent arcs, which we believe to be the most accurate and expeditious known, may be new to some; and this, as well as Prof. Rankine's elegant graphic processes relating to circular arcs, will be found applicable to many other purposes, and exceedingly useful. To which we may add, finally, the hint that the same is true of the principles and the methods made use of in the demonstrations.

C. W. MacCord.

Stevens Institute of Technology,
Hoboken, N. J., Jan. 28, 1881.

PART I.

THE TEETH OF WHEELS.

GENERAL PRINCIPLES.

The proper action of many pieces of mechanism depends so largely upon that of spur wheels, that any means of effecting a radical improvement in the making of such wheels cannot but be of interest and importance.

There was a time when the teeth of wheels were made in rude haphazard ways, of almost any shapes that would permit them to engage, with a mistaken idea that they would wear themselves into correct forms. The machine was expected not only to do its own proper work, but partly to finish itself; small wonder, then, that it failed to do either of these things well. Naturally, these crude methods gave place to better ones. The mechanician perceived the necessity of greater care in making the teeth of proper form ; the mathematician soon became interested in the problem of determining what forms were proper, and the results of their combined efforts, leave little to be desired in relation to the latter.

And as little would seem to be left in regard to the former after the introduction of the gear-cutting engine, by which, if the milling cutter be of the correct outline, all the teeth of a wheel are made with the utmost regularity and precision. But on closer consideration, it will be seen that something is yet lacking in reference to the formation of the cutter itself.

It is one thing to know what its outline should be, but quite another thing to make it so.

The process most extensively employed involves, 1st, the laying out of the required curve; 2nd, the filing of a template to that exact form, and, 3rd, the turning of the cutter to fit the template.

In some cases a specially-contrived apparatus has been used for mechanically tracing the curve by continuous motion, but, until recently, the two remaining steps have been executed by hand, which makes the perfectly accurate formation of a cutter, especially if it be a small one, very difficult, and its exact duplication still more so.

The time has now come when all this ought to be changed. No one who considers for a moment the vast numbers of accurate machines employed in the various industrial arts, and of others equally accurate, employed in making them, can fail to perceive the advantages over the system above described, of one in which the template is not merely lined out, but cut out to the true form, and the contour of the milling cutter, be it large or small, is made to correspond to that of the template, by mechanism nearly automatic. Of such a system, and of the means by which these results are effected, we propose to give a detailed description. Before entering upon this, however, we shall briefly explain the principles upon which the correct forms of the teeth depend, and the method of laying out epicycloidal teeth in outside gear.

GRAPHIC REPRESENTATION OF MOTION.

The motion of a point at any instant may be represented in magnitude and direction by a right line.

It is true that the path of the point may be a curve of any kind; but at any given instant it can occupy but one position, and its direction will be that of the tangent to the path at that point. The length of that tan-gent may, evidently, be made to indicate the velocity; therefore the motion is fully represented.

COMPOSITION AND RESOLUTION OF MOTION.

The *composition* (or finding the resultant) of two motions, is effected as shown in Fig. 1. Suppose the point A, to receive simultaneously two

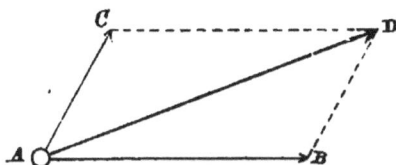

Fig. 1.

impulses, the motions due to which are represented in velocity and direc-tion by AB, AC. Draw BD parallel to AC, and CD parallel to AB: then AD, the diagonal of the parallelogram thus formed, will represent the re-sultant motion in both direction and velocity. That is to say, the point A will go to D, in the same time in which it would have reached either B or C, had it received but one of the impulses.

Evidently, if one component and the resultant be known, the other component may be found in a similar manner. If, for instance, we know that AD is the resultant of two components, one of which is AC, draw CD; then the other component must have the direction AB parallel to CD, and its magnitude is found by drawing DB parallel to AC.

The *resolution* of motion is the converse of composition : thus, it is evi-dent that the motion AD in Fig. 1 may be separated into the components

6

from which it was derived, by drawing the parallels DC, AB, in one direction, and DB, AC, in the other; by which the original parallelogram is reconstructed.

But again, AD may be the diagonal of a great number of other parallelograms: from which we see that a given motion may be resolved into two components, having any directions we please to assign.

ANGULAR VELOCITY AND VELOCITY RATIO.

This term *angular velocity* is applied only to circular motion, like that of a wheel revolving on its axis. Every point in the revolving body turns through the same angle in the same time, whatever be its distance from the axis: thus in Fig. 2, the point A goes to D, in the same time that it

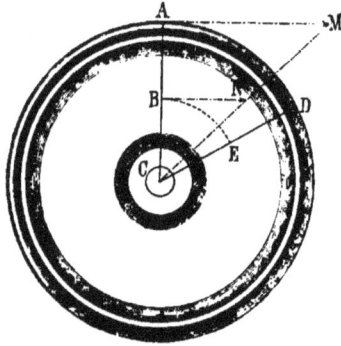

Fig. 2.

takes the point B to reach E. Clearly the arcs AD, BE, represent the *linear* velocities of the moving points; but AD is as many times greater than BE, as AC is greater than BC. If then we divide AD by AC, or

BE by *BC*, either quotient may be taken as the measure of the angular motion represented by *ACD;* or, in general, we say that

$$angular\ velocity = \frac{linear\ velocity}{radius}.$$

If we consider the motion of a revolving point at a single instant only, its direction is that of the tangent to its circular path. In representing it, then, we set off the linear velocity perpendicular to the radius through the point, as *AM, BN.* Drawing *CM,* it will be observed that the angle *ACM* is not the same as *ACD.* Nor should it be, since the former represents what is happening at a given instant, the latter a motion continuing through a period of definite duration.

The *velocity ratio* of two revolving bodies, at any instant, is simply the quotient obtained by dividing the angular velocity of one by that of the other, at the given instant. If this quotient be the same at every instant, the velocity ratio is said to be *constant;* as in the case of two pulleys driven by a belt which does not slip—if one be half the size of the other, it will always turn twice as fast, whether the actual speed be uniform or not.

DETERMINATION OF VELOCITY RATIO.

In Fig. 3, *C* and *D* are fixed centers, about which turn the two curved levers, *CH, DK,* which touch each other at *A.* Through this point draw *TT* the common tangent, and *NN* the common normal, of the two curves, and also *AC* and *AD,* the radii of contact. If *CH* turn in the direction indicated by the arrow, *DK* will be driven in the opposite direction ; we now wish to find the angular velocity ratio of the two motions.

The point A, of the driver CH, must move in a direction perpendicular to CA, and we will suppose its linear velocity at the instant to be represented by AB, which may be resolved into the components AP, AM; the first is the effective component of rotation, the latter being the sliding component.

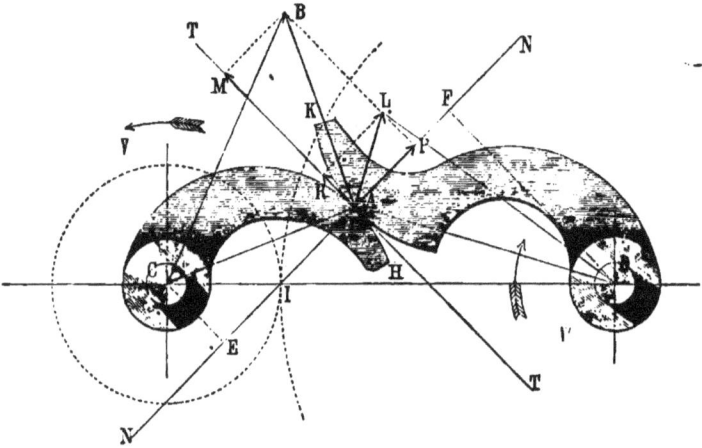

Fig. 3.

The point A, of the follower DK, must go in a direction perpendicular to AD, and its velocity AL must be such as to have the same normal component AP, the tangential or sliding component being AR.

We perceive, then, that when one revolving piece drives another by contact, the motion is transmitted in the direction of the common normal, which is therefore called the *line of action*.

Now draw CE and DF perpendicular to NN, CD the *line of centers*, cutting *NN* at *I*, also *CB*, *DL*. We shall thus have three pairs of similar triangles, viz.:

$$CAE \text{ is similar to } ABP,$$
$$DAF \quad `` \quad `` ALP,$$
$$CIE \quad `` \quad `` DIF.$$

Let the angular velocities of the driver and the follower respectively be represented by v, v'; then since

$$\text{ang. vel.} = \frac{\text{lin. vel.}}{\text{radius}},$$

we shall have

$$v = \frac{AB}{CA}, = \frac{AP}{CE}, .$$

and

$$v' = \frac{AL}{DA}, = \frac{AP}{DF};$$

whence

$$\frac{v}{v'} = \frac{DF}{CE}, = \frac{DI}{CI}.$$

That is to say: *The angular velocities are inversely proportional to the perpendiculars let fall from the centers of motion upon the line of action.*

Or otherwise: *The angular velocities are inversely proportional to the segments into which the line of centers is cut by the line of action.*

CONDITION OF A CONSTANT VELOCITY RATIO.

If it be required that the velocity shall remain *constant*, the second of the values, above deduced, indicates a condition which the curves must satisfy, **viz.**: they must be of such forms that *their common normal shall always cut the line of centers at the same point.* For the line of centers is of fixed length, whence the segments into which it is cut must always be the same, in order to maintain a constant ratio.

Now the outlines of the teeth of spur wheels must be curves which *will* transmit rotation with a constant velocity ratio, and it will readily be seen that parts of the curved levers in the immediate neighborhood of the point of contact *A*, Fig. 3, might be of the forms proper for teeth of wheels whose centers are *C* and *D*. In the figure, the sliding components of the motions *AB*, *AL*, are respectively *AM*, *AR;* these lie in the same direction, but are not equal ; so that at this instant, the levers are sliding upon each other with the velocity *RM*, equal to their difference. There would be no sliding if these were of the same length as well as in the same direction ; but since the normal component *AP*, is the same for each motion, this can only happen when the resultants *AB*, *AL*, also coincide. And these being perpendicular to the radii of contact, cannot coincide unless *CA*, *AD*, lie in one right line, which must, of course, be *CD*. That is to say, there will be more or less sliding, *except at the instant when the point of contact is on the line of centers.*

NATURE OF ROLLING CONTACT.

The finding of curves which will satisfy the above condition, and also the best means of drawing them, depend upon a feature of perfect rolling contact, best seen by a study of that which is not perfect. The polygon, in Fig. 4, rolls along the fixed right line with a hobbling motion ; the point *A* is at rest, and the whole figure turns about it as a center until *B* comes into *LM* at *D*, then about *B*, and so on ; the perimeter of the polygon measuring itself off upon the line. If the number of sides be increased,

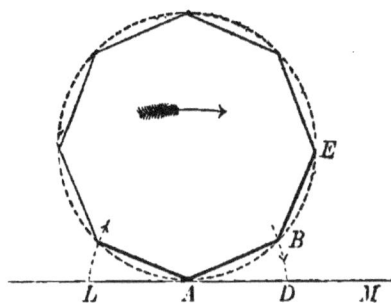

Fig. 4

the hobbling will be diminished, and if the number become inconceivable, it will become imperceptible. The broken outline then becomes the dotted curve, tangent to the line, and the change from one center of rotation to another goes on continuously. But this does not alter the facts, that at any instant the point of contact is *at rest*, and that every point in the figure is at the instant *turning about that point of contact as a fixed center*.

DRAWING OF ROLLED CURVES.

In Fig. 5, AA is a curved ruler fixed to the drawing board, and BB is a free one rolling along it. Let a pencil be fixed to and carried by the latter, either in the contact edge, as at D, or at any distance from it, as at E. At the instant the rulers are in contact at P, the motion of D is in the direction DF, perpendicular to DP, the contact radius. DF, then, is tangent to the path of D, traced as the ruler BB rolls: but it is also tangent to the circular arc whose center is D and radius PD, consequently the path of D is also tangent to that arc. Let the arcs Pc, Po of BB,

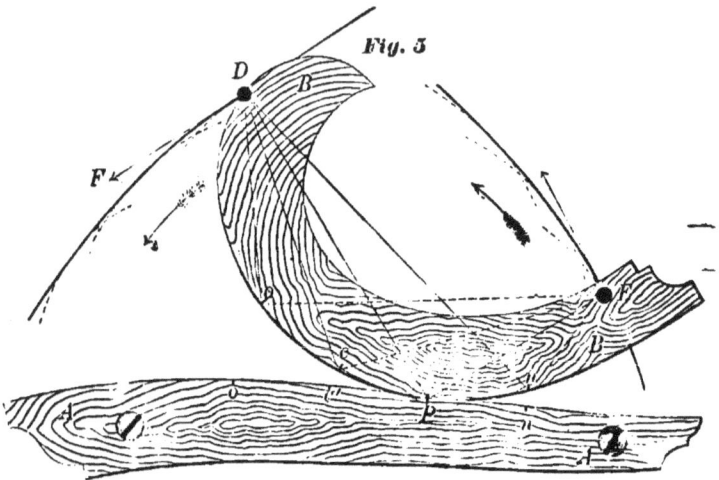

Fig. 5

be equal to the arcs *Pc′*, *Po′* of *AA*, then *cD* will be contact radius when *c* reaches *c′*, and *oD* when *o* reaches *d′*. If, then, we describe, with these radii, circular arcs about *c′* and *o′*, the curve traced by *D* will be tangent to those arcs; and that traced by *E* will be tangent to arcs about the same centers with *cE* and *oE* as radii.

Curves thus described, by points carried by one line which rolls upon another, are called *rolled curves* or *epitrochoids;* and the drawing of a series of tangent arcs as above explained is the readiest and most reliable method of laying them out. The line which carries the tracing point is called the *describing line,* and the one in contact with which it rolls is called the *base line;* either of these may be straight, or both may be curved.

13

RECTIFICATION OF CIRCULAR ARCS.

For our present purpose we have to do only with the rolling of a circle,
either upon its tangent or upon another circle, and shall have frequent
occasion to set off upon a right line a length equal to that of a given arc,
or upon a given circle an arc equal in length to a given right line.

Since the circumference is 3.1416 times the diameter, these operations
can be performed arithmetically; but the following graphic process will be
found equally accurate and much more expeditious.

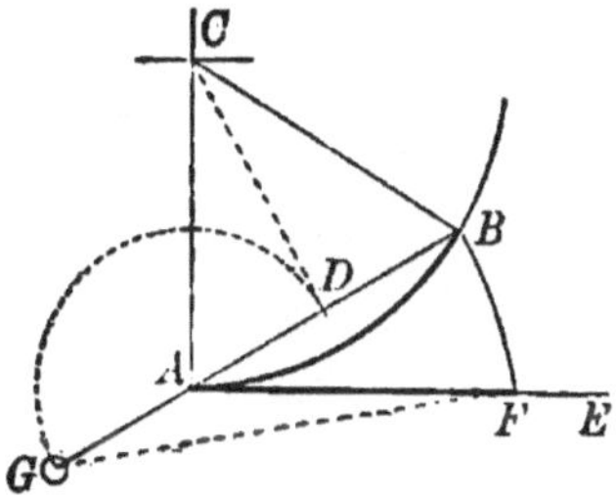

Fig. 6

I. In Fig. 6, let AE be tangent at A to the given arc AB. Draw
BA, produce it, making $AG=AD=\frac{1}{2}$ chord AB. With center G and
radius GB, describe an arc cutting AE in F. Then $AF=$arc AB (very
nearly).

II. In Fig. 7, let the given line AB, be tangent at A, to the given
circle. Make $AD=\frac{1}{4} AB$; with center D, and radius $DB=\frac{3}{4} AB$,
describe an arc cutting the given circle in E. Then arc $AE=AB$ (very
nearly).

NOTE.—The arc thus rectified or found should not measure over 60°.
If the given arc or line exceed this limit, it should be bisected.

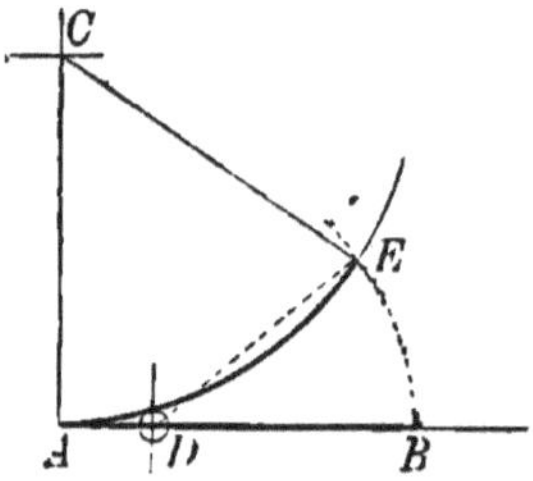

14

Fig. 7

The particular rolled curves to be used are :

I. *The Cycloid*, Fig. 8. Traced by a point in the circumference of a

Fig. 8

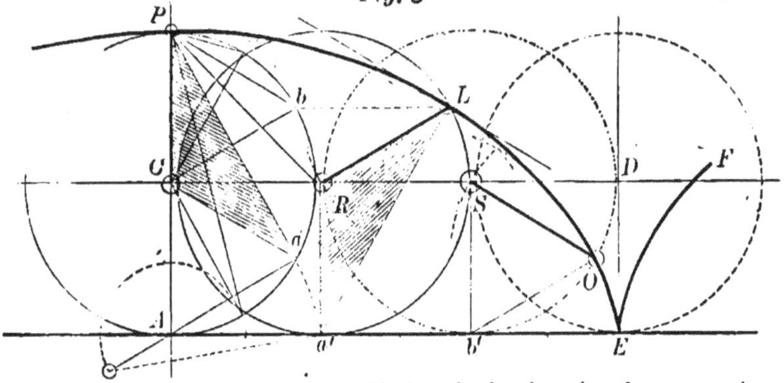

circle rolling upon its tangent. Find *Aa'*, the length of a convenient
fraction *Aa*, of the circumference; step this off the required number of
times, making *AE*=semi-circumference. Divide both into the same

number of equal parts, draw chords from P to the points of division on the circle, with which, as radii, strike arcs about the corresponding points on AE; the cycloid is tangent to all these arcs.

To find points on the curve.—When aC becomes contact radius, it has the position of $a'R$, perpendicular to AE. The angle aCP remaining unchanged, make $a'RL$ equal to it: then RL is the *generating radius*, and L a point on the cycloid. Also $a'L$ is the normal, and a perpendicular to it is tangent to the curve at L.

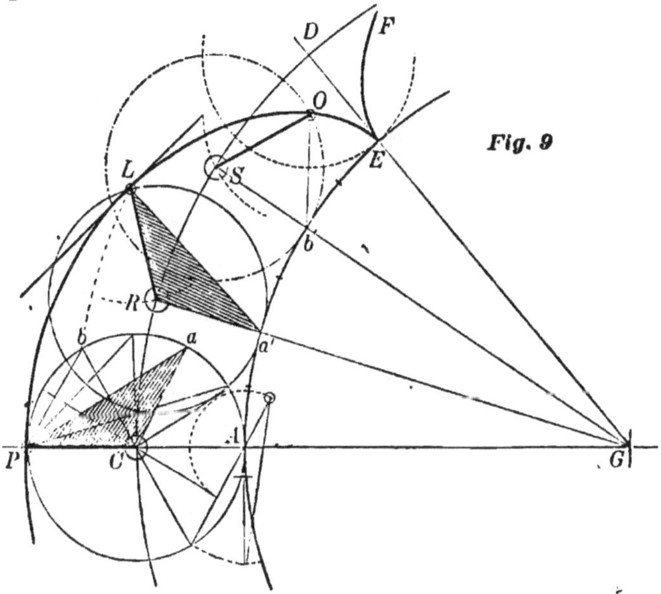

Fig. 9

Conversely. Let O be any point on the curve; about this as a center,

describe an arc with radius equal to *CP*, cutting *CD*, the path of the
center, in *S*. Then *OS* is the generating radius; *Sb'*, perpendicular to
AE, is the contact radius, and *b'O* is normal to the cycloid.

II. *The Epicycloid*, Fig. 9. The describing circle rolls on the *outside*
of another, whose center is *G*. Draw the common tangent at *A*, set off on
this the length of *Aa* (any convenient fraction of semi-circumference *AP*),
and find the arc of the base circle equal to that length. Step this off as
above, making *AE* semi-circumference *AP*. The curve is drawn by
tangent arcs in the same manner as the cycloid. The path of the center
of the describing circle, is, in this case, another circle, whose center is *G*,
and the contact radii *a'R*, *b'S*, are prolongations of the radii *Ga'*, *Gb'*, of
the base circle, which slightly modifies the processes of finding the point
of the curve corresponding to a given point of contact and the converse.

Fig. 10

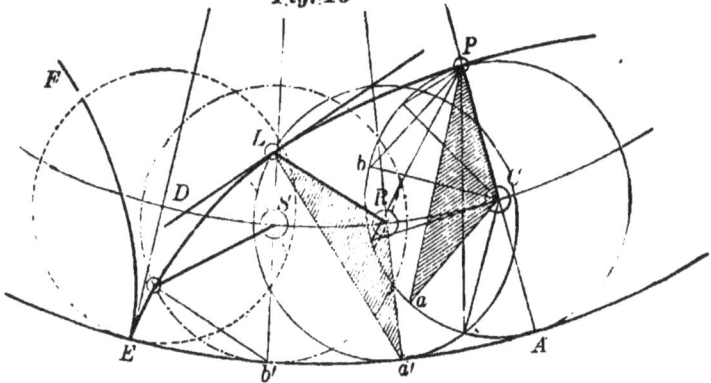

III. *The Hypocycloid*, Fig. 10. Traced by a point in the circumference of a circle rolling *inside* another. Construction in all respects the

same as in the case of the epicycloid, and the diagrams being lettered to correspond throughout, no further explanation is needed.

In all three of these curves, if the rolling continue beyond E, a new branch EF springs up, which is, of course, perfectly symmetrical with EL. It is to be particularly noted that these branches are *tangent* to ED, and to each other, at E. These parts near E are the ones which require the greatest care in their construction, as they only are employed in the formation of teeth.

LAYING OUT THE TEETH — THE PITCH CIRCLE AND CIRCULAR PITCH.

If the line of centers of a pair of spur wheels be divided into two parts which are to each other in the same ratio as the numbers of the teeth, the circles of which these parts are the radii are called the *pitch circles*. And the first step in laying out a pair of wheels is to determine the radii and draw these circles. Suppose, for example, that the distance *CD*, between centers, in Fig. 11, is given, and it is required to make two wheels

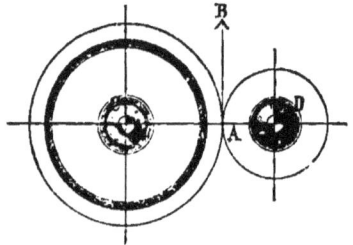

Fig. 11.

whose angular velocities shall be as 2 : 1. Divide *CD* into three equal parts, of which *AD* is one, then *AC* will measure two, and the tangent circles shown are the pitch circles. Evidently they can move in perfect rolling contact about their fixed centers; the linear motion *AB* is the same, whether we regard the point *A*, as belonging to one circle or the other. But one will not drive the other without the possibility of slipping, which would cause the velocity ratio to vary ; hence the necessity of teeth.

The next step is to divide each pitch circle into as many equal parts as its wheel is to have teeth. We may give the smaller wheel any number we please, but the larger one must have twice as many in this instance. The *pitch* of the teeth is the length of the circular arc obtained by this subdivision. Since the larger circumference is twice the smaller, but is divided into twice as many parts, the pitch arc is the same in both wheels. Each of these arcs must contain a tooth and a space; hence we may say that the pitch is the distance between the centers or the corresponding edges of two adjacent teeth, *measured on the pitch circle*, not in a right line. This is sometimes called the *Circular* Pitch, in distinction from what is known as the *Diametral* Pitch, of which hereafter.

GENERATION OF THE TOOTH OUTLINE.

In Fig. 12, let C and D be the centers of the pitch circles *LM, RN.*

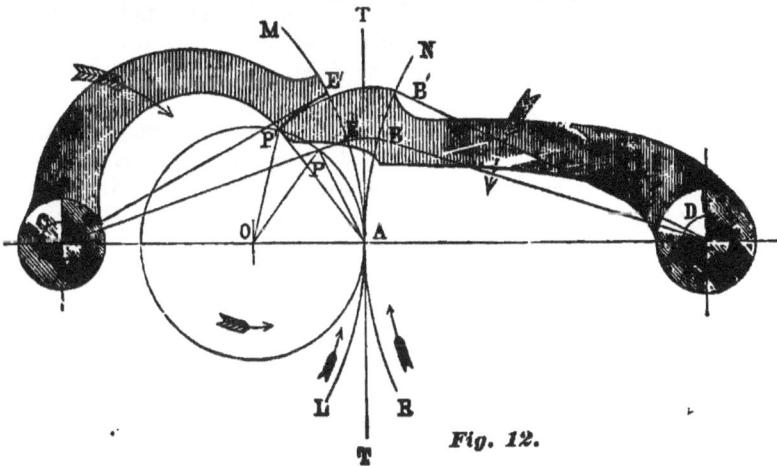

Fig. 12.

Tangent to these at A, is a smaller circle whose center is O. Suppose all the centers to be fixed, then the three circles can move in rolling contact, with equal linear velocities. Set off from A the three equal arcs, AB, AE, AP. Suppose a marking point originally fixed at A, in the circumference of the small circle; then while this travels to P, it must trace, with reference to RN, the curve BP, and with reference to LM, the curve EP. Now the relative motions of the circles are precisely the same as though the small circle had rolled upon the outside of RN, and upon the inside of LM, regarding these as fixed base lines; the curves are, therefore, an epicycloid and a hypocycloid respectively; and AP is their common normal.

If the tracing point go on to P', the arcs AP', AE', AB', being equal, the resulting curves $B'P'$, $E'P'$, are clearly but extensions of the first pair, and AP', is their common normal.

We perceive, then, that the curves thus simultaneously generated are tangent to each other at some point throughout the generation; that the point of tangency is always in the describing circle; and that the common normal always passes through the fixed point A, upon the line of centers.

Consequently these curves are correct outlines for parts, at least, of teeth; if the curved lever CE' turn, as shown by the dotted arrow, it will drive the other before it, the point of contact following the arc $P'PA$, until E' and B' meet at A; and as the common normal always cuts the line of centers at the same point, the velocity ratio will be constant.

FACE AND FLANK.

The epicycloid $B'P'$, which lies without its pitch circle, is called the *face;* and the hypocycloid $E'P'$, which lies within its pitch circle, is called the *flank.* Usually, each tooth has both; but wheels can be made, and sometimes used to great advantage, in which one of a pair has faces only, the other only flanks: we will consider this case first.

SIZE OF THE TOOTH.

This depends upon the pitch, for the pitch-arc must contain a tooth and a space, which might be exactly equal, were perfection in workmanship possible. Practically, the space must be a little wider than the tooth; the difference is called *back-lash*, and should be made as small as practicable. In drawing we may disregard it, and make the thickness of the tooth just half the pitch.

ARC AND ANGLE OF ACTION.

The angle through which a wheel turns, while one of its teeth is in contact with a tooth of another wheel, is called the *angle of action*, and the arc by which it is measured is the *arc of action*. This latter must evidently be at least equal to the pitch-arc, in order that each tooth may continue in gear until the next one begins to act, and it should be considerably greater.

A PAIR OF WHEELS—LIMITING CASE.

In Fig. 13, the pitch and describing circle being drawn as in Fig. 12, let *AB*, *AE*, be the pitch arcs, and *AP* an equal arc on the describing circle. Then the face for the tooth of *RN*, can not be less than *BP*, since, if made, as shown, of exactly that height, contact is ending at *P*, at the very instant the next tooth begins to act at *A*. Bisect *AB* in *H*, which gives the thickness of the tooth ; and draw through *H*, a reversed face similar to *BP*. The conditions are purposely so chosen that this second face

Fig. 13

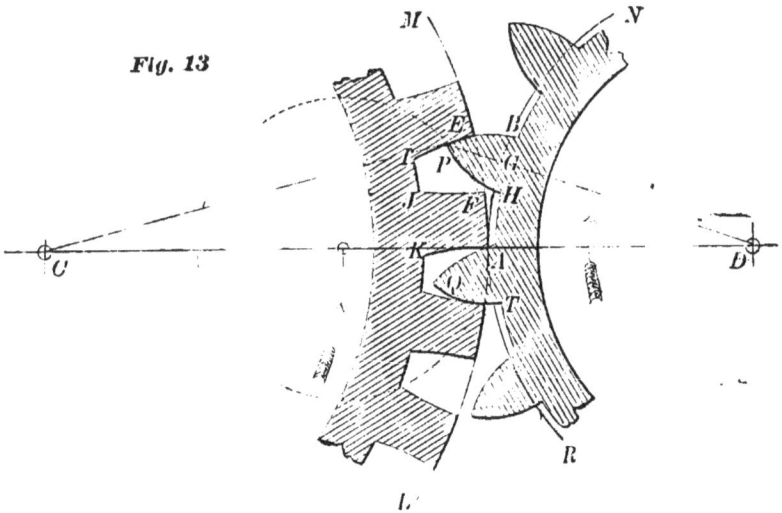

passes through *P*. The case is, therefore, a barely possible one ; the tooth is pointed, and just high enough to continue in gear until the next one begins to act. We found that the face must be of the height *BP*, in order to secure this arc of action; drawing *PD*, which cuts the pitch circle in *G*, we find in this case that *BG* is just half the thickness of the tooth. Had it been greater, *GH* must have been less, so that the face through *H*, would not have passed through *P*, but between *P* and *G*, and the case would have been impracticable ; it would then have been necessary to reduce the pitch and give both wheels more teeth. But if *BG* had been *less* than half the thickness of the tooth, we could either make the tooth higher, or give it some thickness at the top, as in Fig. 14.

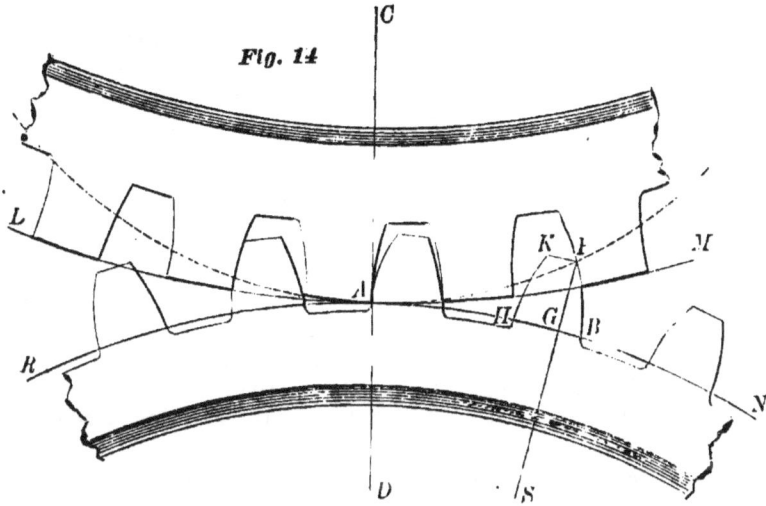

Fig. 14

The *acting* flank is *EP;* but in order to let the teeth of the other wheel pass, the hypocycloid is extended to *I*, making the depth of the space a little greater than *PG;* the difference is called *clearance*, and a similar provision is made in the other wheel by cutting in radially, as shown at *A, H, B*, a little below the pitch circle. The tooth of *LM*, is completed by bisecting the pitch-arc *AE*, at *F*, and drawing the curves *AK'*, *FJ*, etc., similar to *EI.*

A PRACTICAL CASE.

Limiting cases like the preceding are to be avoided in practice. A pointed tooth is bad, as being weak and liable to wear at the top, and

even if it be not pointed, the angle of action should be greater, as otherwise, the least wear at the top reduces the face below the requisite height,
which affects the velocity ratio. A reasonable case is shown in Fig. 14;
the arc of action is 1½ times the pitch, and drawing the radial line *PS*, we
find *BG* much less than ½ *BH*, thus giving the tooth a thickness *PK*, at
the top.

APPROACHING AND RECEDING ACTION.

In Figs. 13 and 14, the action takes place wholly on one side of the
line of centers. If *RN* be the driver (the direction being as shown by the
arrows) the action begins at *A* and ends at *P*, the point of contact contin-
ually *receding* from the line of centers; in which case *AB*, *AE*, are called
arcs of recess, or of receding action. If *LM* drive (in the opposite direction) the action begins at *P* ending at *A;* the point of contact is always
approaching the line of centers, and *AB*, *AE*, are then called *arcs of approach*, or of approaching action.

It has been found by experience that the friction is greater and more
injurious in the latter case than in the former; hence when such wheels are
used, the one with faces only should always drive. But even then, there is
one drawback, which will be seen by reference to Fig. 12. The longer the
arc of action, the longer the face of the tooth, and the greater the
obliquity of the line of action, that is, its inclination to the common tangent
TT, of the pitch circles. The pressure as well as the motion is transmitted
in this line, and the greater its inclination to *TT*, the greater will be the
component of pressure in the line of centers, tending to cause friction in
the bearings.

Consequently such wheels are better suited for use in light mechanism
where the teeth can be made small and numerous, and smoothness of action
is important, than for the transmission of heavy pressure.

TEETH WITH BOTH FACES AND FLANKS.

By giving faces and flanks to the teeth of each wheel of a pair, we can secure a given angle of action with shorter faces, consequently with less sliding and less obliquity of action. Also, the action will take place partly before, and partly after, the point of contact reaches the line of centers. If a wheel is both to drive and to follow, the arcs of approach

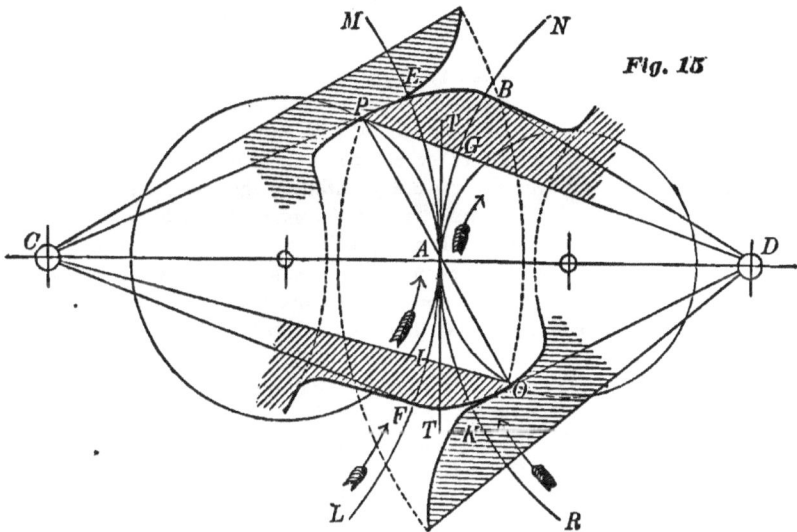

Fig. 15

and recess may be made equal; but if one of a pair is always the driver, it may be desirable to make the arc of recess the greater, in order to reduce the amount of the more injurious friction.

The construction is shown in Fig. 15; all that relates to the face *BP* for *RN*, and the flank *EP* for *LM*, is precisely the same as in Fig. 13, and the lettering being so far made to correspond, no further explanation is needed. To complete the teeth, another describing circle is used, on the opposite side of the pitch circles, which generates the face *OF* for *LM*, and flank *OK* for *RN*. If we assume the arc of action on that side of *CD*, as *AF* or *AK*, the possibility of securing it with a given number of teeth is at once ascertained by making the arc *AO* equal to *AF*, and drawing *OC*, cutting *LM* in *I* : if *FI* be *less* than half the thickness of the tooth as required by the pitch, or *equal* to it, the construction is possible, the tooth in the latter event being pointed ; if greater it is impracticable. If it be found feasible, we have only to draw the epicycloid *OF*, which joined to *EP* completes the outline of the tooth for *LM*, and the hypocycloid *OK*, joining it to *BP*, which finishes the outline of the tooth of *RN*. That is to say, these are the whole of the *acting* outlines ; the flanks must be extended to a greater depth in order to give *clearance*, as already explained.

The operation will be readily seen ; as in the diagram the acting side of a tooth of each wheel is drawn in two positions. Supposing *RN* to drive, the action begins at *O*, the driver's flank pushing the face of the follower, and the point of contact moving in the arc *OA*, until *K* and *F* meet at *A*.

The face of the driver then urges the follower's flank, the point of contact now traveling in the arc *OP*, and at *P* the action ends.

We see, then, that the angle of approach depends upon the length of the follower's face, and the angle of recess upon that of the driver's face ; and if these lengths be assumed or given, those angles are readily found,— as for instance, had the length *FO* been assigned, it is only necessary to strike an arc about *C* with radius *CO*, which, cutting the describing circle in *O*, gives *OA* the length of the arc of approach, which is then to be set off on *LM* and *RN*, as *AF*, *AK*.

A Practicable Example.

The diagram Fig. 15, is drawn without regard to practical proportions, in order to make the construction clear; but in Fig. 16 we have

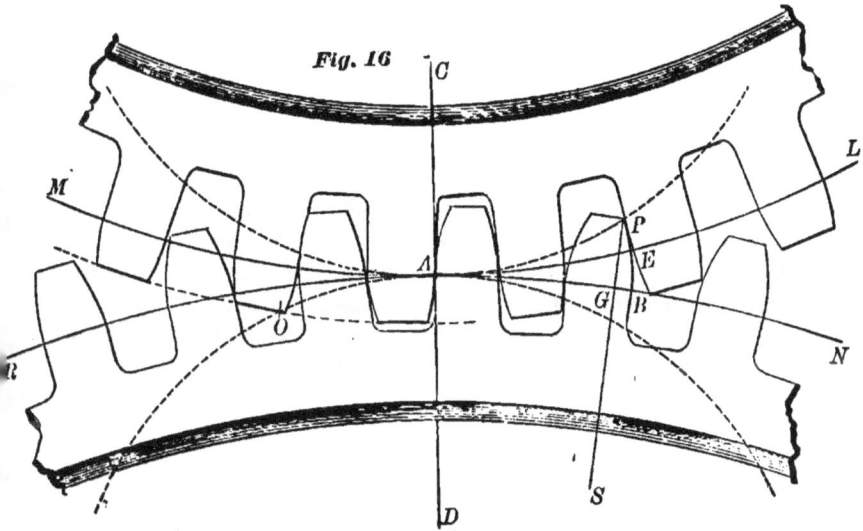

Fig. 16

shown a feasible case; the cut is half size, and the conditions are as
follows:

Distance between centers, 27 inches. Wheels to have 63 and 45 teeth,
the smaller to drive.
Angle of action to be 2⅜ times the pitch.
Angle of recess to be one-third greater than the angle of approach.

28

We have, then, $63:45::7:5$, $7+5=12$, $\frac{12}{2}=2\frac{1}{4}$, $2\frac{1}{4}\times7=15\frac{3}{4}$, $2\frac{1}{4}\times5=11\frac{1}{4}$, for the radii of the pitch circles. And $2\frac{8}{8}=\frac{21}{9}$, which is to be divided into parts in the ratio of $3:4$; whence, $3+4=7$, $\frac{21}{9}\div7=\frac{3}{9}$, $\frac{8}{9}\times3=1\frac{1}{8}$ times the pitch
=angle of approach,
$\frac{3}{4}\times4=1\frac{1}{2}$ times the pitch
=angle of recess.

INTERCHANGEABLE WHEELS.

Inasmuch as the face and the flank which act in contact are generated by the same describing circle, it makes no difference whether the diameter of the one which traces the other face and flank be the same or not, in laying out a single pair of wheels; and in Fig. 15 the describing circles are of different diameters. But for the very reason just stated, it is clear that if we wish to make a number of wheels, any one of which will gear with any other one, we must use the same describing circle for all the faces and all the flanks.

SIZE OF THE DESCRIBING CIRCLE.

In making such a set of wheels, the question at once arises, how large shall the describing circle be? This depends upon a property of the hypocycloid, illustrated by Figs. 17 and 18. In Fig. 17, the describing circle is half the size of the base circle, and the line traced by the point A is merely the diameter AD. For after rolling till the point of contact is B (taken at pleasure), the describing circle cuts AD in P, its center meantime going from E to F. Draw BFC, and PF; then the angle ACB is half the angle PFB, but the radius AC is twice the radius BF, therefore the arcs AB, PB, are equal. In Fig. 18, the point P will trace the curve

Fig. 17

Fig. 18

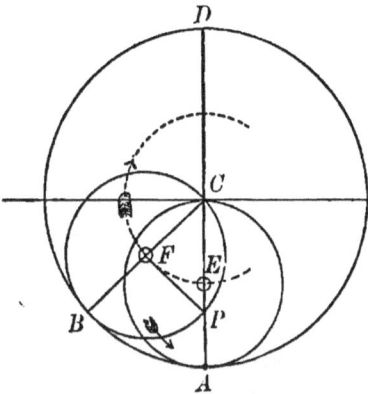

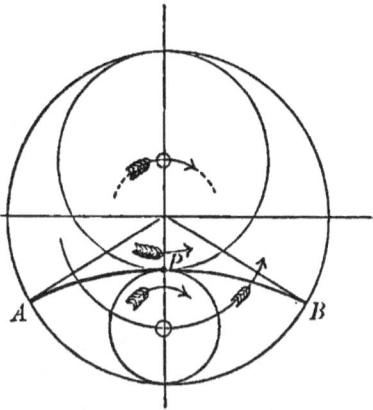

AB, by rolling in one direction, if we regard it as in the circumference of the smaller circle; but if the same point be carried by the larger describing circle, it will trace the same curve by rolling in the other direction. If, then, in any case the describing circle be half the size of the pitch circle, the flanks will be *radial,* as in Fig. 19; if it be less, they will spread out

Fig. 19

Fig. 20

Fig. 21

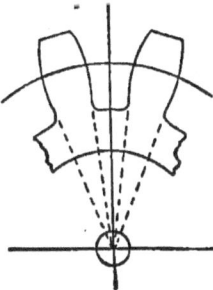

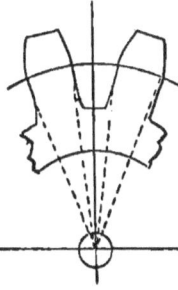

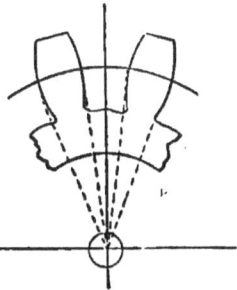

toward the root of the tooth, giving a stronger form, as in Fig. 20 ; but if greater, the flanks will curve in toward each other, as in Fig. 21, whereby the teeth become weaker and difficult to make.

From this, the safe practical deduction is, that the describing circle for a set of wheels should not be more than half the diameter of the smallest wheel ; and in laying out a single pair, two describing circles, each of ⅜ the diameter of its pitch circle, give good practical forms to the teeth for general purposes.

Still, the face is shorter, and the obliquity of action less, for a given arc of action, the larger the describing circle ; so that for very delicate mechanism, it is possible that the gain from these causes would sometimes render it advisable to use teeth of the form shown in Fig. 21. They may be much strengthened by the use of large fillets at the junction of the side and bottom of the space, which is quite admissible, since the *acting* depth of the flank is comparatively small, as has been shown.

RACK AND WHEEL.

A *rack* is simply an infinitely large wheel. The curvature of a circle diminishes as the radius increases, and disappears when the radius becomes infinite ; so that the *pitch line* of a rack is only a straight tangent to the pitch circle of the wheel with which it works, and the line of centers become a perpendicular to this pitch line, through the center of the wheel.

The rack will travel through a distance equal to the circumference of the pitch circle of the wheel during one revolution of the latter, whatever the number of teeth, and in the same proportion for any fraction of a revolution. The pitch of the rack teeth, therefore, is found by rectifying the pitch arc of the wheel, whatever that may be, and setting off that length upon the pitch line.

The construction is shown in Fig. 22. We have here shown the two describing circles as of the same size, and it is clear that if the same circle be used to generate the faces and flanks of a ·set of wheels, any one of them will gear with the rack if the pitch be also the same.

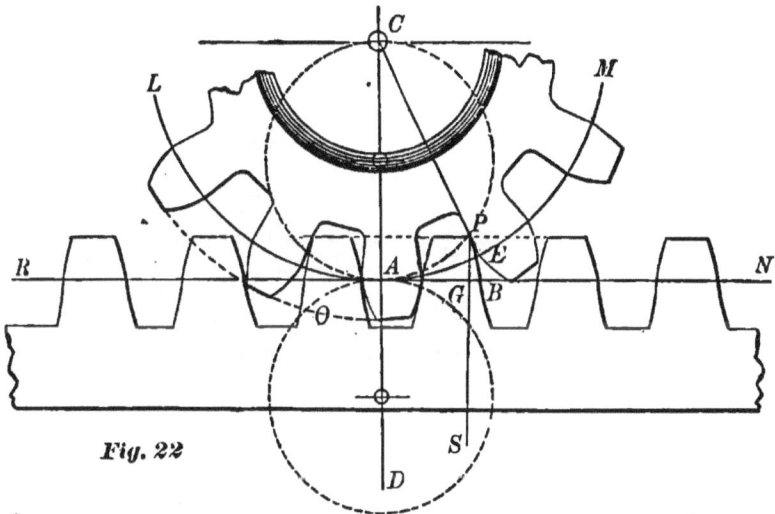

Fig. 22

Evidently both faces and flanks of the rack teeth are cycloids, being generated by the rolling of a circle upon the pitch line. If the length BP of the face, be assumed or given, a line parallel to RN cuts the generating circle in P, thus determining AP, to which AB must be made equal, and fixing the part of the action which will take place on the right of CD. Or if AB be assigned, we make AP equal to it, thus ascertaining the necessary length of face. In either case, PS is now to be drawn perpendicular to the pitch line, which it cuts at *G*, and as in the preceding constructions *BG* cannot be greater, and should be less, than half the thickness of the tooth as determined by the pitch. The part of the action, which will take place on the left of *CD*, depends upon the length of the face of the wheel tooth, and is ascertained as in the cases previously explained.

ARBITRARY PROPORTIONS.

It is not necessary in all cases, indeed probably not in the majority of cases, to pay particular attention to the relative amounts of the approaching and receding action. It is a very common practice to make the whole height of the tooth a certain fraction of the pitch; the part which projects outside the pitch circle being made a little less than that within, by which the clearance is provided for. Thus, in Fig. 23, the whole height

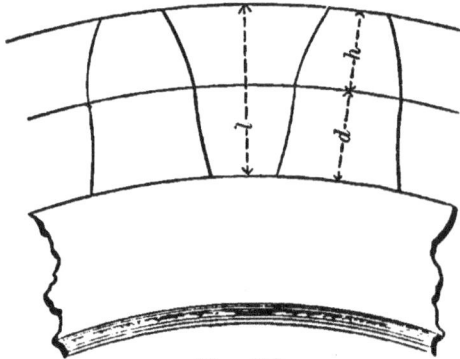

Fig. 23.

l, is ¾ of the pitch; and the part *h* is to *d* as 11 : 13. Two other proportions which have been extensively used are as follows:

$$l = \tfrac{2}{3} \text{ pitch}; \quad h : d :: 4 : 5.$$
$$l = \tfrac{7}{10} \quad `` \quad h : d :: 3 : 4.$$

Teeth proportioned according to either of these rules will work satis-
factorily for most purposes, if there be at least 12 teeth upon the smallest
wheel. But if occasion arises, as it may, for the use of a pinion with only
six or eight "leaves," (as the teeth of very small wheels are sometimes
called,) it will be found advantageous, if not absolutely necessary, to make
the faces of those leaves longer. As for the angle of action, and the proportion of the angle of approach to that of recess, both these things will,
of course, vary according to the numbers of teeth, whichever of these
systems be adopted. It may be added, that in connection with these rules,
instructions are frequently given as to the amount of *back-lash*, which is
also, according to them, a definite fraction of the pitch. This is, no doubt,
very proper in reference to wheels cast from patterns, but there certainly
does not seem to be any reason for it if the teeth are to be cut ; for whatever
the pitch, it is practically sufficient that the backs of the teeth should barely
clear each other when the fronts are in driving contact.

DIAMETRAL PITCH.

In designing spur gearing it is necessary to find the circular pitch, not
only because, as we have seen, it is used in the graphic construction, but
because the strength of the tooth depends upon its thickness. Were there
nothing to the contrary, it would be most convenient to express the pitch
in whole numbers or manageable fractions, as 2 inch pitch, ¾ inch pitch,
and so on. But as the circumference is 3.1416 times the diameter, awk-
ward decimals will often appear in the values of the diameters of the pitch
circles, if this plan be adhered to. Now, if the tooth be strong enough, it
matters not if it be a little stronger ; and it is practically much more important to have the diameter a whole number, or a convenient fraction,
than that the circular pitch should be either the one or the other. This is

accomplished by the use of what is called the *diametral pitch;* which is simply the quotient found by dividing the diameter of the pitch circle, instead of the circumference, into as many equal parts as the wheel has teeth.

Whence, Diametral Pitch $= \dfrac{\text{Diameter}}{\text{No. of Teeth.}}$, and Circular Pitch=Diametral Pitch $\times$ 3.1416.

In the use of this system, convenient values of the diametral pitch are selected, each being a fraction with unity for its numerator and an integer for its denominator, as 1, $\frac{1}{2}$, $\frac{1}{4}$, $\frac{1}{8}$, $\frac{1}{16}$, $\frac{1}{32}$, etc.

The denominators of these fractions only are commonly used in giving the diametral pitch; thus, an "8-pitch wheel" is one which has eight teeth for each inch of diameter, or whose diametral pitch is $\frac{1}{8}''$. This is, in fact, merely inverting the fraction, and giving the value of $\dfrac{\text{No. of Teeth}}{\text{Diameter}}$;

thus, let a wheel of 16 inches diameter have 80 teeth; then, $\frac{16}{80}=\frac{1}{5}''=$ diametral pitch, but $\frac{80}{16}=5$, and we call it a "5-pitch" wheel. By this system the calculations as to diameter and number of teeth are made very simple ; as, for example :

Required, the diameter of a 4-pitch wheel with 37 teeth:

$$\tfrac{37}{4} = 9\tfrac{1}{4} = \text{diameter.}$$

How many teeth of 16-pitch on a wheel of $3\frac{7}{8}$ diameter ?

$$3\tfrac{7}{8} \times 16 = 62 = \text{No. of teeth.}$$

The tooth may be made to project outside the pitch circle, a definite fraction of the diametral pitch, as in the case of the circular pitch; and thus the size of the blank may be readily ascertained. If this projection be made, for instance, 1⅜ times the diametral pitch, the face of the tooth will be nearly as long as that found by the first of the arbitrary rules above given; and the diameter of the blank is determined by simply adding to that of the pitch circle, 2⅓ times the diametral pitch.

PART II.

THE MANUFACTURE OF ACCURATE GEAR CUTTERS.

In cutting a spur wheel it is essential that the contour of the milling cutter conform precisely to that of the space between two teeth. The obvious disadvantages of turning the cutter by hand to fit a template filed out, if not laid out, by hand, have already been alluded to ; also the fact that they have been in part avoided, by the use of mechanical means for describing the required curves on the template.

This is, however, but one step, and that of the shortest; for by the methods previously explained, the epicycloidal curves can be drawn on metal with comparative speed and extreme accuracy. There yet remain the laborious processes of making the template, and making the cutter fit it when made. When these things are done by hand, exact duplication of templates, cutters or wheels, is a matter of impossibility, to all practical intents and purposes.

But this becomes, on the contrary, a matter of ease and perfect certainty, when the work is done by the two machines of which we give illustrations. They were recently introduced by The Pratt and Whitney Company, in whose shops they may be seen in successful operation ; and they are well worthy of study, not only on account of the ingenuity and beauty of their movements, but as models of skillful planning and rare examples of the practical embodiment of correct theory.

37

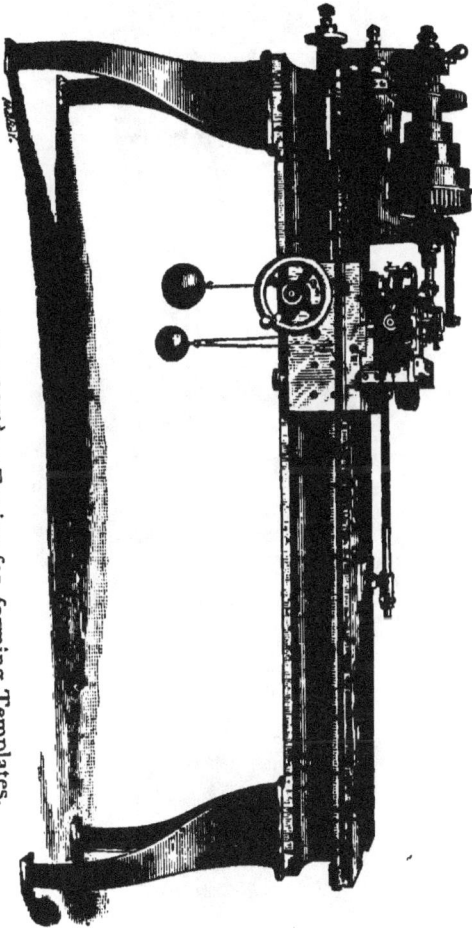

Fig. 1. Epicycloidal Milling Engine for forming Templates.

Fig. 2. Epicycloidal Milling Engine for forming Templates.

THE EPICYCLOIDAL MILLING ENGINE.

The object of this machine is to form the template subsequently used as a guide in shaping the cutter. Its general appearance is shown in Fig. 1; the operating parts are more distinctly shown in Fig. 2, which is a view taken nearly from above; the action is illustrated in Figs. 3, 4, 5, 6, 7, 8.

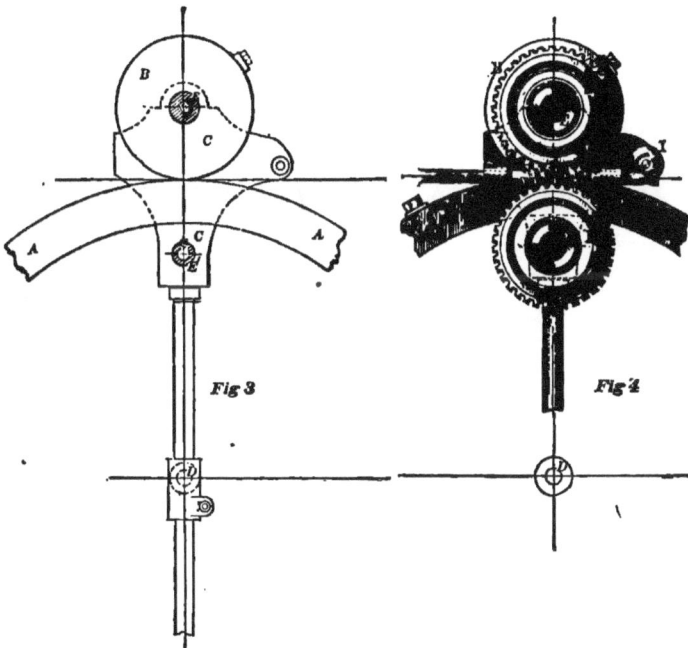

Fig 3   Fig 4

In Fig. 3, *A,A*, is a portion of a flat ring, fixed to the framing; this represents a pitch circle. *B*, is a disc, representing the describing circle;

this turns freely upon a tubular stud E, fixed in the carrier C, which turns about a pivot D, fixed to the frame at the center of A. By means of the clamped socket, capable of sliding upon the rod, the position of D may be adjusted to suit the radius of A. Thus as C moves, the disc can roll upon the edge of A, and is compelled to do so by the flexible steel ribbon shown by the heavy line, which is wrapped round and secured to both pieces, due allowance for its thickness being made in adjusting their radii.

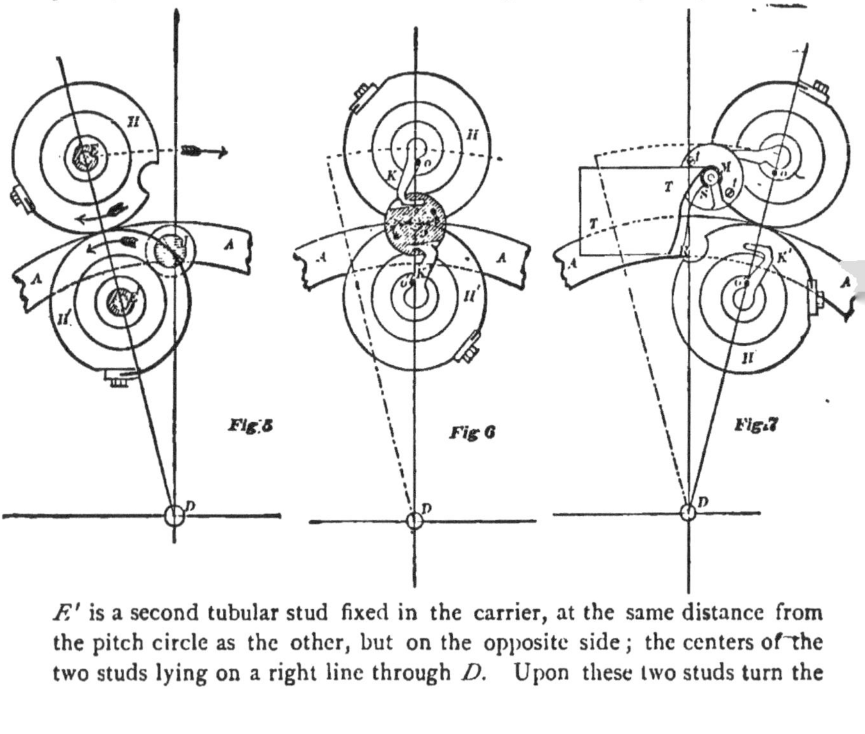

Fig. 5   Fig. 6   Fig. 7

E' is a second tubular stud fixed in the carrier, at the same distance from the pitch circle as the other, but on the opposite side ; the centers of the two studs lying on a right line through D. Upon these two studs turn the two worm wheels, F, F', shown in Fig. 4; these are in a plane above A and B, so that the axis of the worm G, is vertically over the common tangent of the pitch and describing circles; the relative positions of these and other parts will be most clearly seen by a study of the vertical section, Fig. 8. The worm G, is supported in bearings secured to the carrier C, and is driven by another small worm turned by the pulley I, as seen in Fig. 2; the driving cord, passing through suitable guiding pulleys, is kept at uniform tension by a weight, however C moves; this is shown in Figs. 1 and 2.

Upon the same studs, in a plane still higher than the worm-wheels, turn the two discs H, H', Figs. 5, 6, 7. The diameters of these are equal, and precisely the same as those of the describing circles which they represent, with due allowance, again, for the thickness of a steel ribbon, by which these, also, are connected. It will be understood that each of these discs is secured to the worm-wheel below it, and the outer one of these to the disc B, so that as the worm G turns, H and H' are rotated in opposite directions, the motion of H being identical with that of B; this last is a rolling one upon the edge of A, the carrier C with all its attached mechanism moving around D at the same time. Ultimately, then, the motions of H, H', are those of two equal describing circles rolling in ex-ternal and internal contact with a fixed pitch circle.

In the edge of each disc a semi-circular recess is formed, into which is accurately fitted a cylinder J, provided with flanges, between which the discs fit so as to prevent end play. This cylinder is perforated for the passage of the steel ribbon, the sides of the opening, as shown in Fig. 5, having the same curvature as the rims of the discs. Thus when these recesses are opposite each other, as in Fig. 6, the cylinder J fills them both, and the tendency of the steel ribbon is to carry it along with H when C

moves to one side of this position, as in Fig. 7, and along with H' when C moves to the other side, as in Fig. 5.

This action is made positively certain by means of the hooks K K', which catch into recesses formed in the upper flange of J, as seen in Fig. 6. The spindles, with which these hooks turn, extend through the hollow studs, and the coiled springs attached to their lower ends, as seen in Fig. 8, urge the hooks in the directions of their points; their motions being limited by stops o, o', fixed, not in the discs H, H', but in projecting collars on the upper ends of the tubular studs. The action will be readily traced by comparing Fig. 6 with Fig. 7; as C goes to the left, the hook K' is left

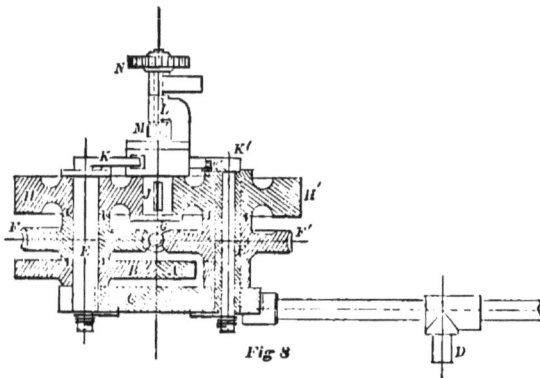

Fig 8

behind, but the other one, K, cannot escape from its engagement with the flange of J; which, accordingly, is carried along with H by the combined action of the hook and the steel ribbon.

On the top of the upper flange of J, is secured a bracket, carrying the bearing of a vertical spindle L, whose center line is a prolongation of that of J itself. This spindle is driven by the spur wheel N, keyed on its upper end, through a flexible train of gearing seen in Fig. 2: at its lower end it carries a small milling cutter M, which shapes the edge of the template T, Fig. 7, firmly clamped to the framing.

When the machine is in operation, a heavy weight seen in Fig. 1, acts to move C about the pivot D, being attached to the carrier by a cord guided by suitably arranged pulleys; this keeps the cutter M up to its work, while the spindle L is independently driven, and the duty left for the worm G to perform, is merely that of controlling the motions of the cutter by the means above described, and regulating their speed.

The center line of the cutter is thus automatically compelled to travel in the path RS, Fig. 7, composed of an epicycloid and a hypocycloid if AA be a segment of a circle as here shown; or of two cycloids, if AA be a straight bar. The radius of the cutter being constant, the edge of the template T is cut to an outline also composed of two curves; since the radius M is small, this outline closely resembles RS, but particular attention is called to the fact that it is *not identical with it, nor yet composed of truly epicycloidal curves of any generation whatever:* the result of which will be subsequently explained.

NUMBER AND SIZES OF TEMPLATES.

With a given pitch, every additional tooth increases the diameter of the wheel, and changes the form of the epicycloid; so that it would appear necessary to have as many different cutters, as there are wheels to be made, of any one pitch.

But the proportional increment, and the actual change of form, due to the addition of one tooth, becomes less as the wheel becomes larger; and the alteration in the outline soon becomes imperceptible. Going still far-

ther, we can presently add more teeth without producing a sensible variation in the contour. That is to say, several wheels can be cut with the
same cutter, without introducing a perceptible error. It is obvions that
this variation in the form, is least near the pitch circle, which is the only
part of the epicycloid made use of; and Prof. Willis many years ago
deduced theoretically, what has since been abundantly proved by practice,
that instead of an infinite number of cutters, 24 are sufficient of one pitch,
for making all wheels, from one with 12 teeth up to a rack.

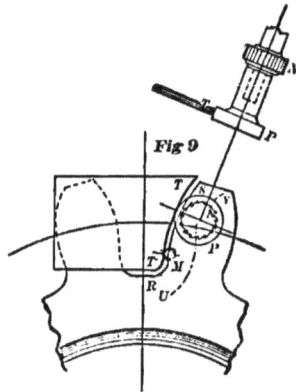

Fig 9

Accordingly, in using the epicycloidal milling engine, for forming the
template, segments of pitch circles are provided of the following diameters
(in inches):

12,	16,	20,	27,	43,	100,
13,	17,	21,	30,	50,	150,
14,	18,	23,	34,	60,	300,
15,	19,	25,	38,	75,	∞.

The diameter of the discs which act as describing circles, is 7½ inches, and that of the milling cutter which shapes the edge of the template, is ¾ of an inch.

Now if we make a set of 1-pitch wheels with the diameters above given, the smallest will have twelve teeth, and the one with fifteen teeth will have radial flanks. The curves will be the same whatever the pitch; but as shown in Fig. 9, the blank should be adjusted in the epicycloidal engine, so that its lower edge shall be ⅟₁₆th of an inch (the radius of the cutter M) above the bottom of the space; also its relation to the side of the proposed tooth should be as here shown. As previously explained, the depth of the space depends upon the pitch. In the system adopted by The Pratt & Whitney Company, the whole height of the tooth is 2⅛ times the diametral pitch, the projection outside the pitch circle being just equal to the pitch, so that diameter of blank = diameter of pitch circle + 2 × diametral pitch.

We have now to show how, from a single set of what may be called 1-pitch templates, complete sets of cutters of the true epicycloidal contour may be made of the same or any less pitch.

THE PANTAGRAPHIC ENGINE FOR FORMING CUTTERS.

In Fig. 9, the edge TT, is shaped by the cutter M, whose center travels in the path RS, therefore these two lines are at a constant normal distance from each other. Let a roller P, of any reasonable diameter, be run along TT, its center will trace the line UV, which is at a constant normal distance from TT, and therefore from RS. Let the normal distance between UV and RS be the radius of another milling cutter N, having the same axis as the roller P, and carried by it, but in a different plane, as shown in the side view; then whatever N cuts will have RS for its contour, if it lie upon the same side of the cutter as the template.

Now if TT be a 1-pitch template as above mentioned, it is clear that N will correctly shape a cutting edge of a gear cutter for a 1-pitch wheel. The same figure, reduced to half size, would correctly represent the formation of a cutter for a 2-pitch wheel of the same number of teeth; if to quarter size, that of a cutter for a 4-pitch wheel, and so on.

But since the actual size and curvature of the contour thus determined, depend upon the dimensions and motion of the cutter N, it will be seen that the same result will practically be accomplished, if these, only, be reduced; the size of the template, the diameter and the path of the roller remaining unchanged.

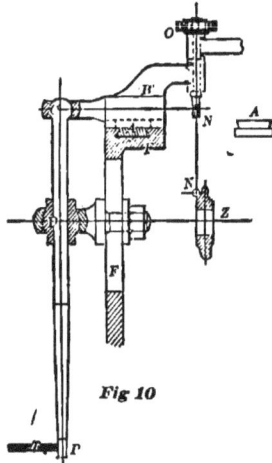

Fig 10

The nature of the means by which this is effected in the Pantagraphic Engine, is illustrated in Fig. 10. The milling cutter N, is driven by a flexible train acting upon the wheel O; its spindle is carried by the bracket B, which can slide from right to left upon the piece A, and this, again, is free to slide in the frame *F*. These two motions are in horizontal planes, and perpendicular to each other.

The upper end of the long lever *PC*, is formed into a ball, working in a socket which is fixed to *B*. Over the cylindrical upper part of this lever slides an accurately fitted sleeve *D*, partly spherical externally, and working in a socket which can be clamped at any height on the frame *F*. The lower end *P*, of this lever being accurately turned, corresponds to the roller *P* in Fig. 9, and is moved along the edge of the template *T*, which is fastened in the frame in an invariable position.

By clamping *D* at various heights, the ratio of the lever arms *PD*, *DC*, may be varied at will, and the axis of *N* made to travel in a path similar to that of the axis of *P*, but as many times smaller as we choose ; and the diameter of *N* is made less than that of *P* in the same proportion.

The template being on the left of the roller, the cutter to be shaped is placed on the right of *N*, as shown in the plan view at *Z*, because the lever reverses the movement.

This arrangement is not mathematically perfect, by reason of the angular vibration of the lever. This is, however, very small, owing to the length of the lever ; it might have been compensated for by the introduction of another universal joint, which would practically have introduced an error greater than the one to be obviated, and it has, with good judgment, been omitted.

The gear cutter is turned nearly to the required form, the notches are cut in it, and the duty of the pantagraphic engine is merely to give the finishing touch to each cutting edge, and give it the correct outline. It is obvious that this machine is in no way connected with, or dependent upon, the epicycloidal engine; but by the use of proper templates it will make cutters for any desired form of tooth; and by its aid exact duplicates may be made in any numbers with the greatest facility. Its general appearance is shown in Fig. 11. It will be noted that the universal joints are not

Fig. 11.   Pantagraphic Engine for forming Cutters.

actually of the ball and socket kind, which suggests the explanation, that in Figs. 3–10 inclusive, we have made no attempt to give precise details or proportions, but only to make as clear as we are able to, the principles and mode of action of these remarkably ingenious machines, as well as of the system adopted in using them.

THEORETICAL DEFECTS OF THE SYSTEM.

It forms no part of our plan to represent as perfect that which is not so, and there are one or two facts, which at first thought might seem serious objections to the adoption of the epicycloidal system. These are:

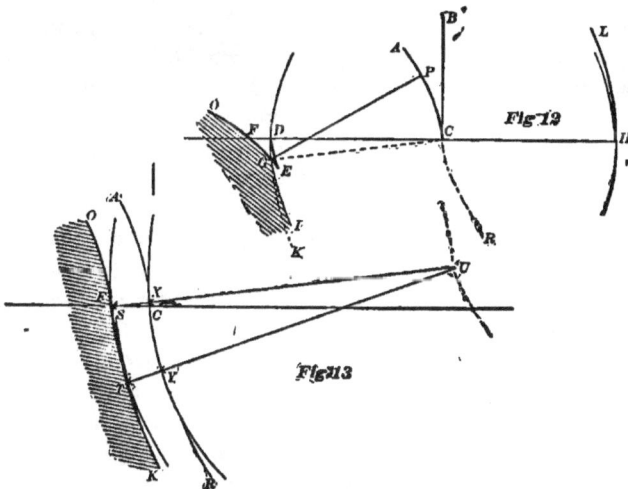

Fig. 12

Fig. 13

1. It is physically impossible to mill out a *concave* cycloid, by any means whatever, because at the pitch line its radius of curvature is zero, and a milling cutter must have a sensible diameter.

2. It is impossible to mill out even a *convex* cycloid or epicycloid, by the means and in the manner above described.

This is on account of a hitherto unnoticed peculiarity of the curve at a constant normal distance from the cycloid. In order to show this clearly, we have, in Fig. 12, enormously exaggerated the radius CD, of the milling cutter (M of Figs. 7 and 8). The outer curve HL, evidently, could be milled out by the cutter, whose center travels in the cycloid CA; it resembles the cycloid somewhat in form, and presents no remarkable features. But the inner one is quite different; it starts at D, and at first goes down, *inside the circle whose radius is* CD, forms a cusp at E, then begins to rise, crossing this circle at G, and the base line at F. It will be seen, then, that if the center of the cutter travel in the cycloid AC, its edge will cut away the part GED, leaving the template of the form OGI. Now if a roller of the same radius CD, be rolled along this edge, its center will travel in the cycloid from A, to the point P, where a normal from G, cuts it; then the roller will turn upon G as a fulcrum, and its center will travel from P to C, in a circular arc whose radius to $GP = CD$.

That is to say, even a roller of the same size as the original milling cutter, will not retrace completely the cycloidal path in which the cutter traveled.

Now in making a rack template, the cutter, after reaching C, travels in the reversed cycloid CR, its left-hand edge, therefore, milling out a curve DK, similar to HL. This curve lies wholly *outside* the circle DI, and therefore cuts OG at a point between F and G, but very near to G. This point of intersection is marked S in Fig. 13, where the actual form of the template OSK is shown. The roller which is run along this template, is *larger*, as has been explained, than the milling cutter. When the point of contact reaches S (which is so near to G that they practically coincide), this roller cannot now swing about S through an angle so great as $\widehat{PGC}$ of Fig. 12; because at the root D, the radius of curvature of DK is only equal to that of the cutter, and *G* and *S* are so near the root that the curvature of *SK*, near the latter point, is greater than that of the roller. Consequently there must be some point *U* in the path of the center of the roller, such, that when the center reaches it, the circumference will pass through *S*, and be also tangent to *SK*. Let *T* be the point of tangency; draw *SU* and *TU*, cutting the cycloidal path *AR* in *X* and *Y*. Then, *UY* being the radius of the new milling cutter (corresponding to *N* of Fig. 9), it is clear that in the outline of the gear cutter shaped by it, the circular arc *XY* will be substituted for the true cycloid.

THE SYSTEM PRACTICALLY PERFECT.

The above defects undeniably exist; now, what do they amount to? The diagrams, Figs. 12 and 13, are drawn purposely with these sources of error greatly exaggerated, in order to make their nature apparent and their

FIG. 14. SET OF WHEELS AND RACK.

existence sensible. The diameters used in practice, as previously stated, are: describing circle, 7½ inches; cutter for shaping template, ½ of an inch; roller used against edge of template, 1⅛ inches; cutter for shaping a 1-pitch gear cutter, 1 inch.

With these data the writer has found that the *total length* of the arc
XY of Fig. 13, which appears instead of the cycloid in the outline of a
cutter for a 1-pitch rack, is less than 0.0175 inch; the real *deviation* from
the true form, obviously, must be much less than that. It need hardly be
stated that the effect upon the velocity ratio of an error so minute, and in
that part of the contour, is so extremely small as to defy detection. And
the best proof of the practical perfection of this system of making epicycloidal teeth is found in the smoothness and precision with which the wheels
run; a set of them is shown in gear in Fig. 14, the rack gearing as
accurately with the largest as with the smallest. To which is to be added,
finally, that objection taken, on whatever grounds, to the epicycloidal form
of tooth, has no bearing upon the method above described of producing
duplicate cutters for teeth of any form, which the pantagraphic engine will
make with the same facility and exactness, if furnished with the proper
templates.

Table of Cutters for Teeth of Gear Wheels,

—MADE BY—

THE PRATT & WHITNEY CO., HARTFORD, CONN., U. S. A

All Gears of the same pitch cut with our Cutters are perfectly interchangeable.

Diameter of Cutters.	Diametral Pitch.	Price of Cutters.	Size of Hole in Cutters.	SET OF 24 CUTTERS. For each pitch coarser than 10.
5 inches.	1½	$25 00	1¼ inches.	No. 1 cuts 12 T
4¼ "	2	20 00	" "	No. 2 " 13
4 "	2½	18 00	" "	No. 3 " 14
3¾ "	3	15 00	" "	No. 4 " 15
3½ "	3½	12 00	1 "	No. 5 " 16
3¼ "	4	9 00	" "	No. 6 " 17
3⅛ "	5	7 00	" "	No. 7 " 18
3 "	6	6 00	" "	No. 8 " 19
2⅞ "	7	5 00	" "	No. 9 " 20
2¾ "	8	4 50	⅞ "	No. 10 " 21 to 22
2⅝ "	9	4 00	" "	No. 11 " 23 " 24
2½ "	10	3 50	" "	No. 12 " 25 " 26
2⅜ "	12	3 50	" "	No. 13 " 27 " 29
2¼ "	14	3 50	" "	No. 14 " 30 " 33
2⅛ "	16	3 00	" "	No. 15 " 34 " 37
2 "	18	3 00	" "	No. 16 " 38 " 42
1⅞ "	20	3 00	" "	No. 17 " 43 " 49
1 13/16 "	22	3 00	" "	No. 18 " 50 " 59
1¾ "	24	3 00	" "	No. 19 " 60 " 75
1¾ "	26	3 00	" "	No. 20 " 76 " 99
1¾ "	28	3 00	" "	No. 21 " 100 " 149
1¾ "	30	3 00	" "	No. 22 " 150 " 299
1¾ "	32	3 00	" "	No. 23 " 300 Rack.
				No. 24 " Rack.

The Cutters are made for diametrical pitches. By diametrical pitch is meant, the number of teeth per inch in the diameter of the gear at pitch line. Two pitches should always be added to this diameter in preparing a gear for cutting. For example: a gear of 80 teeth, 8 to the inch, diametrical pitch, would be 10 inches on pitch circle, but the gear should be turned 10 2-8 (or ¼). The teeth should always be cut two pitches deep beside clearance.

The Cutters are made for a clearance of 1-16 of the depth of the tooth: example: 8 to the inch has a clearance of 1-64; therefore the tooth should be cut two pitches (1-4) and 1-64 deep. The gears must be set to run with this clearance to give the best results.

In ordering bevel gear cutters, give the diameter of gear at outside pitch line, and number of teeth, also the width of face. For the present all cutters are made to order.